# Technology Innovation, Economic Growth, and Geopolitics Policy Options for South Korea

## A Policy Game

YUN KANG, DAVID A. SHLAPAK, SEOK JOON KWON,
DANIEL GONZALES, NATALIA HENRIQUEZ SANCHEZ,
JOHN J. LEE, DONG HO CHUNG, MIYEON OH,
BRADLEY MARTIN, YOU WON PARK, CHRISTIAN CURRIDEN

Prepared for the Chey Institute for Advanced Studies

RAND NATIONAL SECURITY RESEARCH DIVISION

CHEY | CHEY INSTITUTE FOR ADVANCED STUDIES

For more information on this publication, visit **www.rand.org/t/RRA2969-1**.

### About RAND

The RAND Corporation is a research organization that develops solutions to public policy challenges to help make communities throughout the world safer and more secure, healthier and more prosperous. RAND is nonprofit, nonpartisan, and committed to the public interest. To learn more about RAND, visit www.rand.org.

### Research Integrity

Our mission to help improve policy and decisionmaking through research and analysis is enabled through our core values of quality and objectivity and our unwavering commitment to the highest level of integrity and ethical behavior. To help ensure our research and analysis are rigorous, objective, and nonpartisan, we subject our research publications to a robust and exacting quality-assurance process; avoid both the appearance and reality of financial and other conflicts of interest through staff training, project screening, and a policy of mandatory disclosure; and pursue transparency in our research engagements through our commitment to the open publication of our research findings and recommendations, disclosure of the source of funding of published research, and policies to ensure intellectual independence. For more information, visit www.rand.org/about/principles.

RAND's publications do not necessarily reflect the opinions of its research clients and sponsors.

Published by the RAND Corporation, Santa Monica, Calif.


Library of Congress Cataloging-in-Publication Data is available for this publication.
ISBN: 978-1-9774-1-3963

*Cover: Jae Young Ju/Getty Images.*

# About This Report

In this report, we describe a game that was designed and executed by RAND and the Chey Institute for Advanced Studies. The topic was potential policy choices for South Korea that emerge from its central role in the global semiconductor supply chain and its ambitions to occupy a similar position regarding electric vehicle (EV) batteries in an environment that is characterized by increasing U.S.-China competition. The game was executed in August 2023 by participants from diverse backgrounds including security, foreign relations, technology, and commerce.

## RAND National Security Research Division

This research was conducted within the Acquisition and Technology Policy Program of the RAND National Security Research Division (NSRD). NSRD conducts research and analysis for the Office of the Secretary of Defense, the U.S. Intelligence Community, the U.S. State Department, allied foreign governments, and foundations.

For more information on the RAND Acquisition and Technology Policy Program, see www.rand.org/nsrd/atp or contact the director (contact information is provided on the webpage).

## Chey Institute for Advanced Studies

This research was sponsored by the Chey Institute for Advanced Studies, which undertakes policy-relevant research to foster domestic, regional, and international environments conducive to peace and stability on the Korean Peninsula. Chey Institute provides a knowledge-sharing platform that enables cooperation with universities, institutes, and think tanks to analyze various geopolitical risks surrounding the Korean Peninsula and explore opportunities and obstacles posed by scientific innovation.

## Acknowledgments

This report is a product of collaboration among the authors. We express our gratitude to the Chey Institute for Advanced Studies for the funding that made this project possible. We appreciate substantive reviews and inputs made by Lucy Shearer at RAND, Jessica Lin at RAND, Soo Kim at LMI, Naoko Aoki at RAND, and Scott Harold at RAND, as well as Woowon Lee, Min Seung Kim, and Jun Ho Kang at the Chey Institute. We also appreciate the formal substantive reviews performed by Sarah Harting at RAND and Andrew Scobell from the United States Institute of Peace. Lastly, the authors would like to pay special thanks to Ambassador In-kook Park, the former president of the Chey Institute, for his visionary leadership and indispensable support for this project.

# Contents

# Tables

CHAPTER 1

# Introduction

## Background and Motivation

The competition between the United States and the People's Republic of China (PRC) has been rapidly evolving. Although security issues remain at the heart of Washington and Beijing's ongoing contest for global influence, their security competition has extended into economic areas. The principal driver of this evolution has been U.S. efforts to limit the rate of China's progress in developing technology of important military value. Key to the U.S. strategy has been a series of sanctioning actions by which Washington—with the cooperation of such allies as the Republic of Korea (ROK)—has tried to interdict Beijing's ability to import semiconductor manufacturing equipment (SME) and advanced computing chips (U.S. Department of Commerce, 2023).

Semiconductors are central to the modern global economy and serve as the foundation of electronics, powering virtually every aspect of modern technologies, such as consumer electronics and industrial automation. The United States and its allies have enjoyed a substantial—and profitable—position as global leaders in designing and producing chips and chip-making equipment. From a purely economic point of view, if China were to catch up with or perhaps even, in some areas, surpass the technology levels achieved by today's market leaders, this could prove very costly to the United States, Taiwan, Japan, South Korea, and many members of the European Union.

Chips are not only critical to current and future economic success; they are also critical for modern militaries. Current and future military hardware is and will be full of chips. Advanced chips enable the processing of vast amounts of data at the speed and scale required for critical warfighting functions—including communication, command and control (C2), surveillance and reconnaissance, targeting, electronic warfare, and cyber operations—and make the modern military forces operate with far greater lethality, precision, and agility. Military uses of potential breakthrough technologies, such as artificial intelligence (AI), depend on access to the newest generations of semiconductors.

The combined economic and military implications of China making rapid strides toward becoming a global semiconductor superpower make the semiconductor industry a new arena for U.S.-China competition. South Korea faces challenging choices as this great-power rivalry intensifies. South Korea is a global leader in the design and production of critical advanced technologies and aspires to duplicate that success in the burgeoning electric vehicle (EV) marketplace. However, South Korea's policy options are complicated by the simultaneous needs to remain closely tied to the United States and to maintain a generally nonhostile relationship with China. The ROK faces the challenges of finding workable solutions that balance short-term technological, economic, and geopolitical equities with long-term ones.

For more details on the current landscape of the semiconductor industry and geopolitics of semiconductors, see the appendix to this report.

## Game Objective

Our analytic policy game explored a potential future in which the great-power competition between the United States and China intensifies and Beijing continues its pursuit of cutting-edge capabilities in key technology sectors, such as semiconductors. The United States takes steps to decouple itself from Chinese firms in the U.S. military and strategic domestic supply chains and exerts increasing pressure on other countries to do the same. In response, China takes countermeasures and threatens political and trade consequences for countries that acquiesce to U.S. demands.

This game focused on South Korea's perspectives and policies and sought to inform the roles of and implications for government, industry, academic institutions, and other stakeholders.

## About RAND's 360° Gaming Methodology

RAND and the Chey Institute for Advanced Studies executed an analytical policy game—specifically, a 360° Game—that was designed to characterize, analyze, and understand the complex dynamics that govern individuals, organizations, systems, and nation-states. The 360° Game was first developed by RAND in 2015 and has been used by numerous organizations to investigate diverse and and complex policy problems (see Henry, Berner, and Shlapak, 2017).

Unlike a traditional military wargame in which players are in direct competition with one another, the 360° Game provides a setting in which players from multiple disciplines and professions are brought together to share information, generate ideas, and work collectively toward solving a problem (rather than creating problems for one another). The interactive nature of a 360° Game illuminates issues that have been previously overlooked or not well understood and draws important insights that inform stakeholders and decisionmakers.

360° Game participants interact in functional breakout teams that are facilitated by professional research staff. Each functional team, which typically comprises seven to ten individuals, searches for solutions to a specific aspect of the game's overall target problem and report its proposed solutions to a designated senior official. The policy game explores multiple contextual scenarios (or alternative futures), which allows for triangulated analysis that contributes to a 360-degree investigation of both the problem and candidate solutions.

## Report Organization

In Chapter 2, we start by describing a potential future and the geopolitical environment of its global semiconductor industry and then explain the two scenarios that were presented to the game players. In that chapter, we also present the game design and execution, including the players, team organization, and the processes used for gameplay. In Chapter 3, we present the key insights that were derived from the postgame analysis and the implications for South Korea's policy options. The appendix provides the research that the study team conducted on the background information for four areas pertinent to the game—the semiconductor industry, the global competitive environment for semiconductors, the geopolitics of semiconductors, and the EV battery supply chain.

CHAPTER 2

# Gameplay

## Policy Game Background

The game began by introducing the players to a fictional, potential future that was intended as a plausible extension of the current environment, in which the strategic competition between the United States and China intensifies and creates complex technological, economic, and political challenges. The game was executed in August 2023, and although the hypothetical scene-setting portrayed in this chapter closely mimics the real-life environment that existed at the time of gameplay, it was presented as a plausible, potential future that was intended to set up the basis for the participants' discussions exploring various policy options for South Korea.

Geopolitically, the United States has increasingly sought to persuade its friends and allies around the world—especially in Asia—to cooperate in countering China's growing strategic influence. As a result, Japan's defense policies are expanding beyond its postwar "self-defense only" model to embrace expanded power projection in the context of collective security operations (Johnstone, 2022). Australia is planning to invest 368 billion Australian dollars over 30 years to buy and operate nuclear-powered attack submarines (Hurst and Borger, 2023). Even the traditionally reluctant Philippines has granted a new level of expanded base access to the U.S. military under the two countries' Enhanced Defense Cooperation Agreement (Agreement Between the Republic of the Philippines and the Government of the United States on Enhanced Defense Cooperation, 2014).

The United States is also seeking to de-risk its economic relationship with China by revamping supply chains to reduce dependence. Senior officials in the U.S. National Security Council and Department of the Treasury have reiterated that the United States has no intention of pursuing a disastrous decoupling of its economy from China, and further, that "China's economic growth need not be incompatible with U.S. economic leadership" (U.S. Department of the Treasury, 2023).

However, the conflictual dynamics of trans-Pacific U.S.-China competition have been obvious in the high-tech industry. Both the Trump and Biden administrations have imposed ever-tightening limitations on Chinese firms' ability to acquire cutting-edge microchips and the advanced machinery needed to manufacture them from not only U.S. companies but also anyone using U.S.-developed technology in their products. These limits have led major industry actors, such as the Netherlands-based ASML—the maker of the world's most-advanced semiconductor lithography equipment—to halt sales of their most-sophisticated items in China (Deutsch et al., 2022).

The U.S. government has also made its boldest steps into industrial policy in decades to limit both China's progress and onshore manufacturing of high-tech items, the production of which has long been outsourced to Asian countries, including China.

Two pieces of legislation—the Inflation Reduction Act (IRA) and the Creating Helpful Incentives to Produce Semiconductors (CHIPS) and Science Act—are designed to create incentives for high-tech investment in the United States by foreign and domestic companies. At the same time, the legislation requires recipients of federal funds to limit future investment in and sourcing from China (Pub. L. 117-169, 2022; Pub. L. 117-167, 2022).

These actions have had notable effects on China. In the first quarter of 2023, U.S. imports of goods from China were down more than 25 percent year-on-year from 2022 (U.S. Census Bureau, undated). Apple backtracked on a plan to put Chinese-made memory chips in iPhones for sale in China (Ting-Fang, Li, and Yu, 2022). China's leading chip company, Semiconductor Manufacturing International Corporation (SMIC), is projected to grow at only half the previously expected rate in 2023 ("US Tech Curbs Could Halve Growth of China's Top Chipmaker," 2022). Meanwhile, the manufacturers of advanced EV batteries find themselves scrambling to source materials and components outside China.

The ramifications also extend beyond China's borders. The largest Asian semiconductor makers—Taiwan Semiconductor Manufacturing Corporation (TSMC) and Korea's Samsung Electronics and SK Hynix—all have substantial manufacturing presences in China. For example, Samsung Electronics operates the world's largest fabrication facility (fab) for flash memory—also called *not and* (NAND) memory—in Xi'an. Roughly 15 percent of the world's NAND memory is made at the plant (Kim, 2022). SK Hynix operates two plants—in Wuxi and Dalian—which account for close to half of its production of dynamic random-access memory (DRAM) chips ("SK Hynix Says Has Developed Its Most Advanced 238-Layer Storage Chip," 2022).[1] To remain eligible for billions in dollars of U.S. subsidies, all three firms, among others, will need to cut back on plans for expanding or updating factories in China.

In response to U.S. actions, China has accelerated several strategies to enhance its domestic semiconductor industry and reduce its dependence on foreign suppliers, which includes significant investments in research and development (R&D) that are carried out under its Made in China 2025 initiative. China has also sought to strengthen collaboration with domestic semiconductor companies and to attract talent from abroad to bolster its technological capabilities. Additionally, China turned its attention to the leading semiconductor companies abroad, including TSMC, Samsung Electronics, and SK Hynix, seeking their cooperation to advance its domestic semiconductor capabilities. Although China has not yet taken hostile actions that are targeted specifically at these Asian companies, the firms anticipate that such actions might be in the realm of possibility if they continue to resist cooperation with the PRC.

All these concerns were exacerbated in 2023 by a sharp downturn in the semiconductor market, especially in the highly commodified memory segments. Even though the leading memory chipmakers, including Samsung Electronics and SK Hynix, are expected to return to profitability in 2024, the market slump in 2022 to 2023 is a reminder of the volatility of their semiconductor businesses (Chiang, 2023).

There are also concerns about the future of South Korea's overall trade relationship with China. The ROK's exports to the PRC have dropped as a proportion of its global trade ("South Korea Exports Suffer Longest Losing Streak in Three Years, China Sales Tumble," 2023). However, China remained the largest destination for South Korean goods in 2022, account-

[1] Refer to the appendix for background information on NAND and DRAM chips, as well as other semiconductor technologies.

ing for nearly $156 billion in exports, while imports reached $154.6 billion (Jobst, 2023). In other words, China continues to be South Korea's largest trade partner. A further worry is the growth of South Korea's overall trade deficit, which reached approximately $47.8 billion in 2022, the largest since the start of official records in 1956 (Jobst, 2024).[2] This trade deficit marks a steep decline from its record trade surplus of more than $95 billion in 2017 and $29.3 billion in 2021 (Jobst, 2024). Therefore, South Korea has important stakes in the deepening competition between the United States and China. This exercise was intended to explore these stakes.

## Scenarios

After providing an overview of this background, we presented the players with two scenarios that are associated with challenges representing plausible extensions of recent events. The first scenario expands the current challenge: The United States steps up its measures against China and increases its indirect pressures on South Korea. In the second scenario, which was intended to be more complex and challenging than the first, China offers incentives to South Korea to encourage elevated cooperation between Seoul and Beijing in the semiconductor industry. All quotations in these scenarios are fictional.

### Scenario I. Expansion of U.S. Restrictions

In summer 2025, the U.S. presidential administration has become increasingly concerned about developments regarding China. The *New York Times* publishes the explosive results of a months-long investigation that reveals a far wider reaching and more successful Chinese campaign of industrial espionage than had hitherto been uncovered. Experts quoted in the article claim that the industrial espionage campaign, which was targeted globally and broadly, could greatly accelerate China's progress in many technological areas.

---

[2] In 2023, South Korea also recorded a trade deficit with China—$18 billion—for the first time in 31 years because the ROK's exports to the PRC declined almost 25 percent from 2022 (Snyder and Byun, 2024).

The results of this decades-long effort are becoming increasingly visible in China's military posture. The People's Liberation Army (PLA) begins deploying a new version of its Dongfeng-41 (DF-41) intercontinental ballistic missile into dozens of silos that were constructed in China's desert regions. Media reports, which cite U.S. intelligence sources, claim that a new version of the DF-41, the DF-41GH, can carry up to eight highly accurate warheads. Some experts say that the new missile points toward a Chinese desire to develop a true nuclear warfighting capability.

There are also reports in Chinese media and among Internet PLA-watchers, both within and outside China, that a new generation of nuclear-powered attack submarines, the Type 97, is about to begin construction. This Type 97 is the first Chinese sub to fully exploit stolen technology, and the design is rumored to be far quieter than any previous People's Liberation Army Navy (PLAN) submarine, feature better sensors, and have a "plug" of vertical-launch tubes similar to that on the most modern U.S. nuclear-powered submarines. This last feature is especially of interest to U.S. national security because it would give the PLA a subsurface land-attack capability that it previously lacked.

Washington invests increasing amounts in the U.S. military and works urgently with its allies and partners to increase their preparedness against possible Chinese adventurism. Seeking to punish Beijing short of war, the U.S. administration realizes that it has few remaining points of leverage. Having limited options, the United States decides to prohibit the sale of advanced random-access memory (RAM) and NAND memory to a much-expanded group of Chinese firms.

The ruling also requires companies that use U.S. intellectual property (IP) to cease production of the specified RAM and NAND chips at fabs in China. At the same time, the U.S. administration applies substantial pressure on leading chipmakers—including TSMC, Samsung Electronics, SK Hynix, Micron, and Intel—to wind down their operations in China in favor of increased production elsewhere, such as in the United States.

The United States imposes these measures with only the barest of consultation with allies, and all objections are, for all intents and purposes, overridden. South Korean firms, which are the world's largest makers of cutting-edge RAM and NAND chips, are the most severely impacted by the ban. Samsung Electronics and SK Hynix both protest that plans for future investments will be severely disrupted if they comply. Conversations between senior ROK and U.S. officials go nowhere, because U.S. officials argue that the regional security implications of further choking off China's semiconductor enterprise far outweigh any temporary industrial dislocation. U.S. officials also disclose that the U.S. administration is preparing to submit a CHIPS and Science Act II to Congress that will offer tens of billions in subsidies to chipmakers that comply with all U.S. measures against China and that agree to build or upgrade facilities in the United States.

In response, China lashes out against what Chinese officials call "another attempt by the United States to preserve its technological hegemony." China announces new restrictions on the operations of U.S. firms, including limits on the repatriation of profits. Chinese netizens demand the nationalization of the property belonging to firms that comply with the new U.S. rules.

Beijing specifically singles out Samsung Electronics and SK Hynix. According to Chinese officials, both companies have "reaped many billions in profit off the sweat of Chinese labor" and "owe" China in return. Compliance with the new U.S. demands will "be met with a punishing retaliation that these companies will long remember and regret," according to a spokesperson from the PRC.

## Scenario II. China's Carrots and Sticks for EV Batteries

In 2024 and 2025, the United States continues to seek ways of reducing the dependence of the global semiconductor supply chain on China. Working with major U.S.-based information technology firms (such as Amazon, Apple, Google, Intel, Micron, and Nvidia) the U.S. administration lays out a long-term investment strategy that is intended to reshore production capacity outside China, including in the United States.

The governments and technology firms of U.S. allied countries, including Taiwan, South Korea, Japan, India, and European Union states, are brought into discussions about a multilateral Semiconductor Technology Independence Initiative that would, in essence, expand the existing Quadrilateral Security Dialogue (or Quad) into a global partnership that limits sales to and investments in China.

To countermove this initiative, Xi Jinping takes the opportunity at a G20 summit to announce "a cooperative technology development partnership." With $20 billion in initial funding from the China Integrated Circuit Industry Investment Fund (or Big Fund), the partnership is intended to "democratize technology R&D and production" to "overcome certain countries' interference" in global supply chains. Xi claims that "only a truly international win-win partnership" can counter "monopolistic practices" and allow rapid progress. Semiconductors are to be the initial focus of the Chinese partnership, but other high-tech areas may also be targeted.

Although the financial support from the Big Fund is made available to all G20 partners who are willing to join the partnerships, China views South Korea as a critical target for this partnership. South Korean firms operate at the leading edges of semiconductor technology and have already invested billions of dollars in building fabs and other facilities in China. Furthermore, the scale of total trade between the PRC and ROK, especially South Korea's reliance on the Chinese market for its chip exports, appears to have convinced Beijing that it has sufficient leverage to extract increased levels of technology transfer from such firms as Samsung Electronics and SK Hynix.

In bilateral trade talks, China wields both carrots and sticks in an attempt to win South Korean cooperation. Chinese negotiators suggest creating a "common market" among partners for high performance batteries as a "second lane" to incentivize cooperation beyond semiconductors.

In exchange for South Korean firms' enhanced cooperation on semiconductor technology, China offers them preferential access to battery material and components for EVs; both sectors are dominated by Chinese companies. Negotiators also offer expanded access to the Chinese EV market—which is, by, far the world's largest market—to South Korean automakers.

The stick is implicitly brandished when China announces severe restrictions on the export of refined battery materials to the United States. This move of "economic self-defense was necessitated by the unprovoked and seemingly endless" hostility of the United States toward China, according to a spokesperson from Beijing.

As a result, the South Korean government finds itself in a dilemma. Chip exports to China represent 60 percent of South Korea's total semiconductor exports. Overall trade with China is about 20 percent of South Korea's global trade, and any serious disruption would be very costly.

Chip sales into China, which have gradually increased over the past 13 months, are especially important because memory manufacturers, such as Samsung Electronics and SK Hynix, are still recovering from the consequences of the chip glut in 2022 and 2023.

At the same time, the growing sophistication of China's internal semiconductor industry has positioned its own firms, especially Yangtze Memory Technologies Corp (YMTC), to begin challenging South Korea's dominance of the RAM and NAND markets. Paradoxically, the increased fabrication of leading-edge memory chips in China may be necessary for Samsung Electronics and SK Hynix to remain competitive.

Because the highest end of the RAM and NAND markets is likely to become increasingly commodified, South Korean firms become increasingly committed to staking out a leadership position in the burgeoning marketplace for EV batteries. Investment firms pour trillions of South Korean won into the EV battery industry. New technologies and greater capacity gradually increase the global market share of South Korea's three largest EV battery firms—LG Energy Solutions, Samsung SDI, and SK On—to nearly 30 percent.

However, this growth results in South Korean firms' increasing dependence on China for refined materials and battery components, and imports continue to grow. Guaranteed access to steady supplies of these essential materials is necessary for South Korea's EV battery industry's continued progress.

China's exports of EVs continue to skyrocket from less than 1.5 percent of the global market in 2017 to almost 20 percent. This success story worries South Korea's EV manufacturers, which have seen their relative markets in North America and Europe stagnate. For them, the promise of easier access into China's EV market is clearly attractive.

The United States has clearly signaled that it expects South Korea to not cooperate with China in any way that encourages the PRC's progress in any advanced technology. However, South Korean decisionmakers must balance their worries regarding Washington's displeasure with concerns regarding the consequences of China's retaliation if Seoul were not to cooperate, at least in some measure.

## Players

The 360° Game is designed to encourage creative discussion among participants that have different backgrounds and competing stakes in the problem under consideration. For this exercise, we included individuals who have backgrounds in government, technology development, business and industry, trade and economics, foreign relations, and security and who would have to work together to arrive at a final decision on a policy matter. The diversity of the players' backgrounds is important to ensure that balanced perspectives are represented during the discussions.

Fifteen experts from across these disciplines participated in our game. Overall, the group of players represented diverse expertise areas in their professional careers that were distributed fairly evenly across three areas: (1) economics, trade, and finance; (2) security and diplomatic relations; and (3) technology. The group also covered multiple sectors of government, private corporations, and academia or research institutions (which included think tanks). About half of the players were affiliated with institutions in South Korea and the other half with U.S. institutions at the time of the game. All of the U.S.-affiliated players had significant experience in working on issues related to South Korea or the Asia-Pacific region in one or more of the mentioned expertise areas. Table 2.1 shows the profile of the game players.

The discussion was held under the Chatham House Rule; the players were promised anonymity in exchange for their frank and honest participation. English was the language used throughout the event; all the players were fluent in English, including the experts from South Korea, so no interpreters were used.

## Team Organization

All players were told that they had been recruited to serve on a high-level South Korean government commission that has been tasked with assessing how Seoul's policies should respond to each of the two scenarios. Because the commission's remit spanned multiple ministries, this commission would report to the prime minister on behalf of the president.

**TABLE 2.1**
**Profile of the 15 Total Game Players**

| Characteristic | Number of Participants |
|---|---|
| Expertise | |
| Economics, trade, and finance | 8 |
| Security and diplomatic relations | 6 |
| Technology | 6 |
| Sector | |
| Government | 6 |
| Corporation | 6 |
| Academia or research institution | 11 |
| Country of Affiliation of Work | |
| South Korea | 8 |
| United States | 7 |

NOTE: For Expertise and Sector, the sum of the number of participants is greater than the total number of players (15) because some participants had significant career experience in more than one area.

After each scenario was presented, the players were split into two teams; each team represented a different perspective on the problem presented in the scenarios.

One team's focus was on *security and diplomatic relationships.* This team was instructed to approach the scenario from the perspective of avoiding or limiting damage to key relationships—including with China and the United States—and expanding South Korea's diplomatic influence in Northeast Asia and more broadly in Asia.

The other team's focus was on *economic growth and innovation.* This team was instructed to address each scenario with the objective of sustaining the health of the South Korean economy and promoting its position as a leading-edge player in such important areas as technology and finance.

In accordance with the 360° Game construct, each team had players from multiple disciplines in all three areas (i.e., science and technology; economics, trade, and finance; and national security and diplomacy) who were experts from various sectors—public, private, and academia.

## Game Execution

The game was conducted during a full-day event at RAND's Washington, D.C., office. The event began with a plenary session in which the full group of players were presented with the game background described earlier, and then the first scenario was introduced. The players had the opportunity to question and comment on the scenario to ensure that they understood the policy problem they were supposed to address.

The players were divided into their assigned teams and moved to different rooms. For about 90 minutes, they engaged in discussions on multiple topics—understanding the policy problem, identifying the stakeholders' challenges and equities, and assessing a variety of potential policy solutions—and these conversations were moderated and coordinated by a RAND facilitator.[3] Specifically, the players were asked to examine the challenges that were presented in the first scenario by considering the following guiding questions:

- Which South Korean interests are most at risk in this situation?
- What policy actions should Seoul take to mitigate consequences?
- What are the short- and long-term risks of those steps?
- What do you recommend that the ROK do?

At the end of the breakout session, the teams reconvened for a second plenary discussion, during which a representative from each team presented a synopsis of their discussion and their top recommendations. The prime minister, who was portrayed by a senior RAND researcher, and members of the other team had opportunities to challenge the presenting team with questions.

The process was repeated for the second scenario with the following guiding questions:

[3] The RAND facilitator was a senior researcher who has decades of in-depth experience in core areas related to this study, including game design and execution and international security in Europe and in the Asia-Pacific and other regions.

- How closely can South Korea cooperate with China before jeopardizing its security relationship with the United States?
- What are the economic consequences of increased cooperation with China?
- Would South Korea's high-tech development benefit from more cooperation with China, and if so, how much?
- How can fallout with the United States and China be managed depending on the ROK's choices?

At the end of the day, the group held a hotwash session, in which the participants were given opportunities to reflect on the key takeaways and lessons learned from the team discussions about the two scenarios.

## Postgame Analysis

During all discussion sessions, including the team breakout sessions, two notetakers kept detailed records.[4] After the game, the Chey Institute for Advanced Studies team worked with us to organize the notes and analyze them to identify several features: the points of agreement and disagreements; the points of ambiguity, if any; the variety of potential policy solutions explored by the teams; and the final recommendations. We paid special attention to how decisions on the recommendations were reached. We then used this analysis to generate a list of key insights to address the policy dilemma on geopolitics, economics, and technology for South Korea.

---

[4] Players remained anonymous and were not identified by name in these notes.

CHAPTER 3

# Game Insights

The game's outcomes and subsequent analysis led to the following findings that capture the challenges in keeping South Korea's technological, economic, and geopolitical equities balanced with one another. Significant caution must be applied when seeking to draw lessons from the single play of a game. Little, if anything, can be said to have been proven. Furthermore, one of the primary products of the game is the collective and individual experience of the participants; in other words, each player walks away with their own lessons from the event, regardless of what lessons are written down afterward. Therefore, our insights in this chapter are formulated on the issues and themes in which we saw a convergence of the players' views from our postgame analysis.

## South Korea's Interest in U.S. Partnership Is Not Easily Put at Risk by Chinese Economic Measures Alone

Across both teams in the two scenarios, there appeared to be a fairly strong consensus that, although South Korea should maintain good relations with China, the ROK's interests clearly demanded a continued close partnership with the United States. Even though participants acknowledged that U.S. policy choices could inflict "shocks" on the ROK's economy, moving closer to China was perceived to carry a "risk premium" that was incommensurate with any expected benefits.

There was a sense, at least among some players, that China is an untrustworthy partner for South Korea. Although erratic U.S. policies from one administration to the next might produce uncertainties, fears of Beijing choosing not to abide by any deals made with the ROK meant—in the eyes of these participants—that closer ROK-China relations were not worth the potential risk to the U.S.-ROK alliance.

It was clear that the carrots offered by China in the second scenario, which appeared quite substantial in the authors' opinion, were inadequate to motivate any meaningful shift in the teams' policy choices. Although most players agreed that none of the Chinese incentives provided in the scenario was worth undermining U.S.-ROK relations, at least one player noted that supply-chain incentives may not be entirely unattractive from an industry perspective. For example, a joint mining operation to secure lithium, nickel, or cobalt would capture the attention of South Korean firms that need these critical EV battery materials. However, such mining operations to secure critical materials by South Korea, acting alone, would require approximately eight to ten years, not to mention a significant amount of funding and investment.

Nonetheless, we observed that a question remained whether any geopolitical incentive from China would prove significant enough for South Korea. For example, one question that was not included during the game design because of time and resource constraints was: Would a credible Chinese offer to broker a serious breakthrough in inter-Korean relations be attractive enough to risk straining the U.S.-ROK alliance?

## Trade-Offs Across Time: Short-Term Perspectives vs. Long-Term Ones

A consistent theme throughout the game was the need to consider trade-offs across time, especially the challenge of managing short-term pains while taking steps to convert these pains into long-term gains.

For example, in the first scenario, the participants claimed that the scope and extent of the new U.S. limits marked a transition from "sanctions" to "economic warfare." The participants assessed that South Korea would be a "loser" in the short term because its semiconductor industry sacrificed sales and reorganized operations to comply with the U.S. restrictions. However, several participants articulated the prospect of the ROK being a winner in the long term because of the disruption to China's emergence as a serious competitor in the high-tech industry. Restrictions on exports to China would force Beijing to return to a more autarkic approach to developing technology. Furthermore, because the PRC is already two (or more) generations behind the world in chipmaking, China's attempt to catch up to market leaders, such as SK Hynix or Samsung Electronics, would be greatly postponed, if not prevented entirely.

The players also noted a caveat that comes with this assessment. Although South Korea may be able to take advantage of China's slowdown in the high-tech foundry business in the long term, the ROK is not entirely free of the PRC's growing competitiveness in the memory chip industry. In particular, the technology gap in NAND manufacturing is less than one generation, which is significantly smaller than the two (or more) generation gaps in DRAM manufacturing. In the short term, closer cooperation between the U.S. government and South Korean firms that operate in China is needed to maintain the ROK's upper hand across the entire memory-chip industry.

Much of our participants' discussions centered on the question of whether—and how—South Korean firms could ride out the short-term damage inflicted by more-severe sanctions on China. A general consensus emerged that some variety of government support via direct subsidies or incentives would be needed. At least one player acknowledged that it was through this sort of "industrial policy" that South Korea and the other *Asian Tigers* had grown their high-tech industries.[1]

If ROK firms with fabs operating in Wuxi, Dalian, and Xi'an were to feel the pressure to gradually downgrade or even relocate their manufacturing capacity outside China, how much support Washington would be willing to provide for these transitions remains a subject of further analysis, according to the participants' discussion; they noted that even though this approach has been absent from U.S. policy, the CHIPS and Science Act indicates a new willingness from Washington to directly intervene in fostering specific techno-industrial economic sectors.

[1] The term *Asian Tigers* refers to the developed economies of Hong Kong, Japan, Singapore, and South Korea, which have achieved a high rate of growth fueled by industrialization and exports since the 1960s.

## The U.S.-China Rivalry Requires South Korea to Keep Its Political, Security, and Economic Interests in Balance with One Another

According to how the scene-setting was done in the game background section (see Chapter 2), the ideal situation for South Korea, similar to many U.S. partners and allies in the Indo-Pacific region, would be to gain maximum benefit from economic interchange with China while retaining a full and reliable security commitment from the United States. For many years, the ROK (and others) have been able to approach this ideal balance and enjoy a space in which they were not forced to take sides between the two great powers.

The participants from both teams agreed that the increasingly antagonistic U.S.-China relationship has rendered this balancing act to be untenable over the long term. Even though no participant suggested that a complete severing of ROK-PRC economic ties is likely, desirable, or even possible, South Korea's relationships with the two sides can no longer remain largely unlinked.

There was no inclination among participants to sacrifice, or even erode, the U.S.-ROK relationship to improve South Korea's ties with China. At the same time, however, our participants acknowledged the damage that the PRC could do economically to the ROK via, for example, restricting the sale of processed EV battery materials and components to South Korean firms that are highly dependent on them.[2] In addition to these risks, the participants discussed the losses that South Korean firms must absorb by complying with the new sourcing and assembly requirements imposed by the IRA on EV batteries and battery materials.

The discussion then turned to what South Korea might need in terms of a new balance point between Washington and Beijing. Trade-offs among the various aspects of the ROK's relationships with the two great powers will grow more acute. For a concrete example from the second scenario in our game, even though the players seemed immune to China's inducements for cooperation, one participant observed that "the short-term disruption

---

[2] However, it was granted that, to use the words of one player, "pulling the trigger" on supply restrictions only works once because customers will invariably seek to diversify supply in response. Furthermore, imposing such restrictions, even once, would result in a more pervasive lack of trust and motivate a desire to find other sources more generally.

to the South Korean economy" by Beijing's threatened retaliation cannot be overstated. Thus, although the larger overall question of which way to lean in the competition appeared unambiguous and in favor of the United States, there was a clear awareness of the costs and risks involved.

It is worth noting that, even though the issue of South Korea's domestic politics did surface in the players' discussion, those considerations were not prominently highlighted in the policy choices presented to the prime minister. The trade-offs featured in those conversations were broadly those trade-offs entailed in the management of South Korea's *external* economic and strategic choices. What those choices are—and the results flowing from them in terms of the well-being of South Korea's people—will enormously influence the conduct of politics in the ROK. This three-way trade space is complex and highly dynamic, and navigating this environment will be very challenging for South Korean leaders.

## What Are the Key Takeaways on Strategy for South Korea?

The postgame analysis of all the conversations that took place across both scenarios in the game identified the key findings presented in the previous sections. Our analysis also offered the following set of coherent strategic approaches for South Korea to address its technological, economic, and geopolitical challenges.

First, even though the relationship between South Korea and the United States is complicated and, at times, problematic for the ROK, this relationship is critical not only to South Korea's security but also to its economic health. The relationship must be kept strong and, when possible, further strengthened.

Second, strengthening the relationship will require changes in South Korean and U.S. behavior. Discussion of the U.S. actions in the scenarios alluded to concerns over parts of the CHIPS and Science Act and the IRA that were widely seen in South Korea as discriminating against ROK firms. According to one participant, the United States can take action to alleviate South Korean (and other allies') concerns by offering "a seat at the table," thereby making sure that any U.S. decision would not lead to unforeseen or unanticipated disadvantages to the private sectors of its allies. In other words, faits accomplis are unwelcome, even among friends.

This proposed action also implies that South Korea, along with Japan and Taiwan, will want to assume more-active roles in discussing future innovation in semiconductor technology with the United States. Such cooperation in developing and setting standards for next-generation chips will benefit both the United States and its allies because they will be able to contribute to different aspects of the semiconductor supply chain.

Third, South Korea should work with the United States and others to agree on a common suite of export controls and to diversify the sources of supply away from China. The Coordinating Committee for Multilateral Export Controls could serve as a good foundational model.[3] Using this multilateral platform, member states could get involved in the early process of devising export controls and reach a common understanding regarding the consequences of leveling such sanctions. This collaboration would afford member states and their industries enough time to minimize damage.

Finally, South Korea must seek to put some sort of "floor" under its relationship with China. Even though the ROK-U.S. relationship remains critical and is unlikely to be put at risk by economic incentives by China, South Korea needs a basis on which it can continue to preserve its ties with China. The premises on which this might be constructed were not specified, but the idea of South Korea serving as a trusted interlocutor or a mediator between China and the United States was mentioned several times. This approach is not without challenges. The United States and China may not welcome South Korea taking on this role, and South Korea could find itself wading in dangerous waters. However, the ROK establishing itself in such a role could ensure that some level of dialogue between Seoul and Beijing can be maintained even while the U.S.-China competition grows more intense.

[3] The Coordinating Committee for Multilateral Export Controls was set up during the Cold War by the United States and its allies to coordinate export controls on the Soviet Union and prevent it from accessing key military technologies.

APPENDIX

# Background Material

This appendix presents the background research that the study team conducted in preparation of the game. The purpose of the research was twofold: (1) to inform the game designers on the key facts and issues in the areas of technology, economics, trade, and geopolitics for the semiconductor and EV battery supply chains, focusing on South Korea and its key trading partners, and (2) to prepare the game participants who may not be in-depth experts in *all* these areas before the event. This appendix was distributed to the participants ahead of the game.

The background information is organized into four areas covered by the policy game:

1. semiconductor industry basics
2. semiconductor competitive environment
3. the geopolitics of semiconductors
4. battery supply chains.

This material presents the pertinent facts and observations that are relevant to the scenarios in the game and was intended to inform the game participants' discussions. This appendix is not meant to be a comprehensive report on the semiconductor or EV battery industries.

## Semiconductor Industry Basics

There are multiple categories of semiconductor devices; this report focuses on the two types that make up two-thirds of the semiconductor market:

logic chips and memory chips.[1] Logic chips are packed with transistors that process digital information using logical or arithmetic operations. Memory chips use tiny transistors to store the digital data that logic chips manipulate.

## Logic Microchips and Major Players

The most common way to compare the capabilities of logic chips is the node size of the chips, which is measured in nanometers (nm). Historically, the term *nm* referred to the size of specific features of the chip's circuitry but has increasingly simply become shorthand for referring to different generations of chips (Hruska, 2021).[2]

Leading-edge logic chips power the world's most-advanced computing applications, including AI, gaming systems, supercomputers, and advanced military systems.

Samsung Electronics and TSMC are the only firms mass-producing chips at the smallest current node, 3nm, as of 2023. Both companies, along with Intel, plan to produce chips with nodes of 2nm and smaller, perhaps as soon as 2024 (Wilson, 2023).

## Memory Chips and Major Players

Memory chips are typically subdivided into two categories: DRAM chips and NAND chips.

### DRAM Chips

DRAM is called *volatile memory* because the data stored in the chip will be lost over time if the chip is not periodically refreshed by an electric current. DRAM memory is fast and is typically paired on a computer motherboard with the central processing unit (CPU)—a type of logic chip—to store program instructions and intermediate results.[3]

---

[1] There are other types of chips used in computer and communication devices, such as analog-to-digital converters, which are specialized and made using different materials and production processes. These types of chips are not considered in this report.

[2] For example, Intel's 10nm manufacturing process produces chips with feature sizes, component densities, and performance that are very similar to what Samsung Electronics and TSMC call a *7nm* process (Hruska, 2021).

[3] Many modern CPUs have substantial amounts of extremely fast RAM built in, which processing units use to store information that is needed the most rapidly or most often.

The performance of DRAM is often measured by the chip's capacity and processing speed. Capacity refers to the amounts of data chips can hold, which are typically measured in gigabits (Gb), or billions of single binary bits of data. Processing speed refers to the time it takes to access or add to information stored on the chip.

Different DRAM designs are optimized for different uses. As of 2021, SK Hynix manufactures the highest-capacity chips available at 24 Gb. This chip has a processing speed of 819 gigabytes (GB) (Song, 2021).[4] As of 2023, SK Hynix is the sole manufacturer of high-bandwidth memory for Nvidia; these DRAM chips are compatible with graphics processing units (GPUs)—a class of logic chips extensively used in AI applications. Such advanced memory chips are becoming critical to enhancing AI capabilities (Choi, 2023).

Samsung Electronics has announced a 24 Gb chip for use with GPUs that processes data at up to 1.1 terabytes (which equals 1 trillion bytes, or 1,000 GB).

Samsung Electronics, SK Hynix, and Micron are the three largest DRAM manufacturers, having 41 percent, 29 percent, and 26 percent shares, respectively, of worldwide revenue as of the third quarter of 2022 (Alsop, 2023b).

### NAND Chips

NAND—also called *flash memory*—is nonvolatile memory that provides long-term data storage. NAND chips are typically used to replace hard-disk drive (HDD) technology for mass, permanent information storage. Universal serial bus (USB) sticks, solid-state drives (SSDs), smartphones, and robotics are common uses for NAND memory. Although NAND is slower than DRAM, NAND memory offers much faster access than HDD and is more resistant to mechanical damage because it has no moving parts.

NAND chips are typically built by stacking layers (i.e., by adding word lines) to maximize the storage capacity for a given die size.[5] As of spring 2023, the leading-edge chips produced by SK Hynix and Micron have 238 and 232 layers, respectively, and both have 512 Gb capacity. Samsung Electronics plans to produce a 1 terabit, 236-layer NAND chip (Mellor, 2022).

---

[4] One byte is typically eight bits.

[5] The term *word line* refers to a horizontal strip of polysilicon that connects to the control gate of a transistor. The term *die size* refers to the physical size of a chip.

Samsung Electronics, Kioxia, and SK Hynix controlled 31 percent, 21 percent, and 19 percent shares, respectively, of the global NAND market by revenue as of the third quarter of 2022 (Alsop, 2023c). Two U.S. companies—Micron and Western Digital—together have about 25 percent of the market. Much of the remaining 6 percent is shared by Nanya Technology and a relatively new competitor, the Chinese semiconductor manufacturer YMTC.

## Semiconductor Supply Chain Makeup

The supply chain for producing the microchips is usually described as having four stages:

1. R&D
2. design
3. fabrication
4. assembly, testing, and packaging (ATP).

The semiconductor supply chain is both globally dispersed and regionally concentrated by stage, which is shown in the following descriptions of each stage.

### Stage 1: R&D

R&D is what drives the continual progress in increasing the speed and capacity of information technology products while lowering their cost. Firms headquartered in the United States accounted for 56 percent of the total $80.5 billion that was invested in R&D worldwide in 2021. Taiwanese companies accounted for 14 percent, followed by South Korea and Japan at 12 percent each ("R&D Spending Dominated by U.S. Chip Manufacturers," 2022).

### Stage 2: Design

In the design stage, the techniques and capabilities derived from R&D are translated into the specifications for a new chip. Major components of this segment include firms that hold IP cores and those that provide electronic design automation (EDA) services. The term *IP cores* refers to reusable bits of a microchip design "with a defined interface and behavior," similar to an instruction set (Platzer, Blevins, and Sutter, 2020, p. 13).

The Japanese-owned, United Kingdom–located firm Arm, Ltd., is the leading provider of IP cores and has a "near-monopoly" of the smartphone industry (Clarke, 2023). However, U.S. firms are estimated to have about 52 percent of global market share, while European firms, whose largest competitors are Intel and AMD, have about 43 percent (Khan, Peterson, and Mann, 2021, p. 8).

## Stage 3: Fabrication

Fabrication entails the actual manufacturing of chips, which is an intricate and complex process that can require hundreds of steps (Platzer, Blevins, and Sutter, 2020, p. 15). Fabrication is a highly capital-intensive enterprise; state-of-the-art production fabs can cost upward of $7 billion (Lewis, 2019, p. 11). Highly specialized equipment is required, and individual machines can weigh 400,000 lb and cost $123 million (ASML, 2017).

We will delve more deeply into the fabrication process—specifically, the sources of raw materials and the importance of various distinctions among chips of different nodes—in the Geopolitics of Semiconductors section of this appendix. Here, we note that among integrated device manufacturers—the companies that design and produce their own chips—Samsung Electronics has the largest market share by revenue at 11 percent, while Intel and SK Hynix are the runners up at 9.7 percent and 6 percent, respectively (Alsop, 2023a).

The so-called pure play (or foundry) sector—which includes the companies that manufacture chips according to the designs of other firms—is becoming increasingly prevalent. This area is dominated by TSMC, which claimed a 58.5-percent market share in 2022 (Alsop, 2023d).

## Stage 4: ATP

ATP is the final stage before distribution to end users. ATP can be performed by the chipmaker but is often outsourced to semiconductor assembly and test firms.

Even though ATP requires its own quota of specialized equipment, this stage is labor-intensive, does not require skills in either design or fabrication, and, thus, represents a lower value segment in the supply chain. Therefore, many companies may offshore their ATP services. Firms based in the United States, Taiwan, China, and South Korea are key players in this segment (Khan, Peterson, and Mann, 2021). As of 2023, Taiwan leads the global ATP market.

## Semiconductor Competitive Environment

The growing importance of semiconductors in every aspect of modern life and advanced technology has inevitably made them targets of economic competition since at least the 1980s, when Japanese firms challenged the long-standing dominance of U.S. chipmakers (for example, see Miller, 2022). The geopolitical dynamic that has arisen from the growing strategic competition between the United States and China has been added to that economic dimension. The decisions that were formerly driven almost exclusively by commercial considerations—such as where to build a new fab or whether to license IP cores to a potential customer—are now complicated by the contest between a U.S. desire to slow China's technological and military advances and Beijing's determination to prevent this. We attempt to discuss these two dimensions separately, but their entanglement makes it impossible to entirely detach one from the other.

### Commercial Competition

The supply side of the semiconductor market is dominated by the United States, South Korea, Taiwan, and Japan, which is indicated by some of the data cited in the previous section. Although firms from these four countries compete with one another in many ways, their products and services are also complementary in many parts of the semiconductor value chain.

In 2022, the United States led the semiconductor industry with a 46-percent market share, which was about the level it has held since the late 1990s. South Korea had a 21-percent slice of the pie, while Japan, Europe, Taiwan, and China had 9 percent, 9 percent, 8 percent, and 7 percent, respectively (Semiconductor Industry Association, 2022, p. 3).[6]

[6] The national statistics cited usually refer to firms headquartered in the specified country. Because of the multipolarity of the semiconductor supply chain, these firms almost always have significant operations in other parts of the world.

However, these shares conceal the significant segmentation along national lines among the phases and components of the supply chain. For example, the United States has only about 12 percent of global semiconductor manufacturing capacity (Varas et al., 2020). However, fabless design firms in the United States, such as Nvidia, Broadcom, and Qualcomm, account for more than two-thirds of global revenue in that segment (Alsop, 2023a). Similarly, two South Korean firms, Samsung Electronics and SK Hynix, control about 70 percent of the global DRAM market and roughly half of the global NAND market (IC Insights, 2022; Nam, 2022). Meanwhile, Taiwan dominates the pure-play (or foundry) category; as noted earlier, TSMC alone generates almost 60 percent of that segment's revenue while South Korea's Samsung Electronics is second with a 16-percent share (Alsop, 2023d).

The highly sophisticated tools that are needed to create chips come from a variety of sources, and firms in the United States, Japan, and Europe monopolize the market. The most-advanced lithography machines—which etch the pattern of circuitry onto silicon wafers—are marvels of modern engineering. The smallest nodes sizes currently available (3–5 nm) or planned (1–23 nm) be achieved only by using extreme ultraviolet wavelengths—which are so small (13.8 nm) that they cannot be focused with conventional lenses and instead require incredibly flat mirrors, such as distributed Bragg reflectors.

As of 2023, only Samsung Electronics and TSMC were mass-producing chips at the 3 nm node (Brown, 2022; Park, 2022). In fact, *all* of the world's capacity at 10 nm or smaller nodes was located in either Taiwan (92 percent) or South Korea (8 percent) (Varas et al., 2021).

This supply chain is truly globalized. Chips designed in the United States (or developed elsewhere using EDA software from U.S. sources, which control over 95 percent of this small but critical market) are fabricated in either Taiwan (specifically, logic chips) or South Korea (specifically, DRAM, NAND, and increasingly logic chips) using advanced technology that mostly comes from Europe, the United States, and Japan (Khan, Peterson, and Mann, 2021, p. 8).

This synergistic arrangement has emerged through the combined benefits of commercial advantage and government policy. The new twist in this story is China's growing role and even greater ambitions in the semiconductor economy, as well as how that role and those ambitions have become entangled in the growing U.S.-China strategic rivalry.

## Strategic Competition

We have discussed the semiconductor industry in terms of the performance of firms *headquartered* in a given country—the United States, South Korea, and so on. However, the impact of U.S.-China competition can be fully described only by discussing where these multinational conglomerates actually *operate*. China matters greatly in this discussion.

We previously described how two South Korean firms—Samsung Electronics and SK Hynix—dominate both the DRAM and NAND memory markets. However, not all of their production takes place in South Korea. For example, Samsung Electronics has the world's largest NAND fab, which is able to produce 250,000 wafer sheets per month, in the centrally located Chinese city of Xi'an (Kim, 2022). SK Hynix also operates in China and runs two plants there, including a DRAM plant in Wuxi that produces 180,000 wafers per month and a NAND plant, which was acquired from Intel, in Dalian ("SK Hynix Says Has Developed Its Most Advanced 238-Layer Storage Chip," 2022). In addition, Both Samsung Electronics and SK Hynix own so-called back-end fabs in China for ATP. Meanwhile, TSMC, the foundry giant, operates a so-called gigafab that uses state-of-the-art 300 mm/12-inch wafers in Nanjing, China (TSMC, undated).[7]

All three firms retain their most-advanced manufacturing capacity in their home countries, but the technology is not confined. Apple is reportedly the sole initial customer for TSMC's 3 nm chips, but Baidu, a Chinese technology company that is often referred to as "China's Google," is among Samsung Electronics' initial customers (Hwang, 2023). Although it is not possible to reverse engineer a chipmaking process, the acquisition of the most-cutting-edge available chips by a Chinese firm is an example of how easily and rapidly very advanced technology can diffuse.

In addition to being a site for fabs owned by overseas firms, China is also a prominent supplier of raw materials for semiconductors. However, except

[7] Intel, the largest U.S. logic chipmaker, operates one production site and two ATP facilities in China (Intel, 2023).

for a small number of specific minerals, this is largely a matter of production economics; for example, other countries have substantial reserves of silicon, but silicon can be extracted more cheaply in China.

Furthermore, similar to the governments of the countries in which the other major players in the semiconductor industry are headquartered, China's government has been promoting the development of an onshore chip industry for years through the provision of incentives that come in various shapes, kinds, and amounts.[8] To date, however, these efforts have not produced the results that Beijing might have expected; according to a 2021 report, other than being a vendor of raw materials, China captures only 6 percent of the global added value of the semiconductor supply chain (Khan, Peterson, and Mann, 2021, p. 8).

Unlike the countries where other semiconductor producers are headquartered, China is believed to have undertaken a systematic and long-term campaign of IP theft. At a 2020 conference, U.S. law enforcement officials said they had identified more than 1,000 cases of Chinese theft of U.S. technology and pegged semiconductors as one of the main areas of IP theft (Cimpanu, 2020). Furthermore, in 2023, ASML—the Dutch manufacturer of the world's most advanced lithography equipment—said an employee in China had misappropriated "data relating to proprietary technology" (Gross, 2023; see also Bhattacharjee, 2023).

## The Geopolitics of Semiconductors

The United States, in particular, views China's campaign to improve its semiconductor technology base dramatically as not only a commercial problem but also a national security challenge. In addition to nearly every other area of modern life, microchips are at the heart of advanced weapon systems. Whether used in the guidance kit for a smart bomb, a sophisticated radio designed to hinder enemy snooping on transmissions, the cameras on spy satellites, or the intelligence fusion processes that inform strategic,

[8] Becoming "a global leader in all segments of the semiconductor industry" has been an articulated goal of the Chinese government since 2014 (Lewis, 2019, p. 2).

operational, and tactical planning and C2, semiconductors are vital to the functioning of modern armed forces.

Because the competition with China has intensified, Washington has sought to slow Chinese technological advances. The United States has imposed increasingly tight controls on the kinds of chips and manufacturing technology that can be sold to Chinese firms as part of this strategy. However, such restrictions bode ill for the countries caught between the two sides. South Korea, which is a treaty ally with the United States and an important economic partner for China, is tasked with the tall order of balancing long-term security interests with emerging economic concerns that cannot be overlooked in the global-digital era.

In May 2019, the Trump administration added China's telecommunications giant Huawei to the Entity List (U.S. Department of Commerce, 2019). The move, which was expanded in August 2020 to cover more Huawei-affiliated firms (including HiSilicon, a fabless subsidiary of Huawei), requires special licenses from the U.S. government for "the export, reexport, or transfer (in-country)" of items to Huawei or its affiliates (Bureau of Industry and Security, U.S. Department of Commerce, 2020, p. 51596). This requirement closed off Huawei's access to U.S. chips and semiconductor technology.[9]

In 2020, China's largest chipmaker, SMIC, was added to the Entity List for items required to produce semiconductors at advanced technology nodes of 10 nm and smaller (U.S. Department of Commerce, 2020).

Then, in June 2021, the Committee on Foreign Investment in the United States (CFIUS), an interagency committee charged with oversight of proposed overseas investments in U.S. firms, blocked a proposed $1.4 billion merger between South Korean chipmaker Magnachip Semiconductor, Ltd., and Chinese private equity firm Wise Road Capital (Picarsic and de La Bruyere, 2022; "U.S. Chipmaker Magnachip, China's Wise Road End $1.4bln Merger Deal," 2021).[10] This blocked merger was seen as surprising because

---

[9] In 2019, the U.S. Department of Commerce also put Fujian Jinhua IC, a Chinese memory chip startup, on the entity list for theft of IP (Barbe and Hunt, 2022, p. 10).

[10] Wise Road Capital is a "government-backed, -guided, and -partnered Chinese investment vehicle" that seeks to "buy up cutting-edge technology all across the global semiconductor supply chain" (Picarsic and de La Bruyere, 2022).

Magnachip had no employees or operations in the United States. However, its parent firm, Magnachip Semiconductor Corporation, is a Delaware-registered holding company and listed on the New York Stock Exchange. These were apparently seen as "touchpoints" that justified CFIUS's intervention (Flicker et al., 2021).

In March 2022, the U.S. government proposed setting up a semiconductor industry alliance, known as the *Chip 4 Alliance*, with its Asian partners—Japan, Taiwan, and South Korea—to keep the Chinese semiconductor industry in check. Whereas Japan and Taiwan have expressed willingness to join the alliance, South Korea's position on the alliance remains ambiguous (Gargeyas, 2022).

In August 2022, the CHIPS and Science Act, which seeks to create investment incentives in the domestic U.S. semiconductor industry, was signed into law. Importantly, the CHIPS and Science Act places restrictions—called *guardrail provisions*—on firms accepting funding under the Act that limit them from investing in the expansion and technological upgrades of semiconductor manufacturing in countries of concern, including China.

Then, in October 2022, the Biden administration placed even stricter controls on China's technology access by exploiting what one analyst called "U.S. dominance" of four "'chokepoint' (or, as it sometimes translated from Chinese, 'stranglehold') technologies in the global semiconductor technology supply chain": high-end AI chips, advanced EDA, advanced SME, and components and SME manufactured by non-U.S. firms using U.S. technology (Allen, 2022, p. 2). As of March 2023, the two other countries with the largest shares of the SME market—Japan and the Netherlands—had instituted similar restrictions on sales of advanced SME (Swanson, 2023).

There is no doubt that China is feeling the consequences of Washington's newest policies; they will be long-lasting and cut across many domains, as the United States intended. For example, SMIC was projected in 2023 to grow half as fast in terms of sales as was previously anticipated, according to analysts' estimates ("US Tech Curbs Could Halve Growth of China's Top Chipmaker," 2022). YMTC, which is China's largest DRAM and NAND chipmaker, had been making significant advances in the sophistication of its products to the point that Apple intended to use YMTC's NAND chips in iPhones made for the Chinese market. The October 2022 restrictions caused the tech giant to backtrack on those plans (Ting-Fang, Li, and Yu, 2022).

Almost certainly in response to sanctions, China has barred U.S. chipmaking giant Micron "from key infrastructure projects" because of "serious network security risks" identified in the company's products (Hoskins, 2023). Meanwhile, China also continues to delve into autarkic approaches to technological advancement. In September 2023, SMIC announced it had independently developed 7 nm logic chips for use in Huawei devices. SMIC's chips have allegedly been fabricated using a multipatterning technology that is based on deep ultraviolet lithography without the use of extreme ultraviolet devices, which are fully restricted from being imported into China by U.S. sanctions (Kirton and Cherney, 2023).

## Effects of U.S. Efforts on South Korea

The South Korean business community appears to be apprehensive about the U.S.-led effort to constrain China's semiconductor progress. In September 2022, Chey Tae-won, the chair of SK Hynix, underscored the importance of ROK-PRC trade links (Lee, 2022). Approximately 60 percent of South Korean–made chips are exported to China, although it should be noted that much of that volume is incorporated into products that are assembled under contract and then re-exported; China itself does not consume over half of South Korean chip production (Lee, 2022). Chey Tae-won further noted that South Korean firms leaving the Chinese market is not an option because of these critical economic ties.

Thus far, the South Korean government has joined meetings to discuss the Chip 4 Alliance but has not decided on its participation in the alliance yet. Seoul has been apprehensive about joining the alliance because of possible economic retaliation from Beijing akin to China's economic coercion following South Korea's Terminal High Altitude Area Defense (THAAD) deployment.

South Korean firms also have concerns over the rules to control funding distributed under the CHIPS and Science Act, which include the following:

- limits on expansion of semiconductor manufacturing capacity for leading-edge and advanced facilities in China to 5 percent of production capacity over a five-year period
- limits on the expansion of legacy semiconductor facilities in China[11]
- restrictions on recipients from engaging in joint research or technology licensing efforts with a foreign entity of concern that relates to a technology or product that raises national security concerns (U.S. Department of Commerce, 2023).

These worries have been somewhat assuaged by the Biden administration granting waivers from the most troublesome provisions to chipmakers in both South Korea and Taiwan (Saxena, 2023).

## Battery Supply Chains

### South Korea's Battery Industry Has World's Major Players and Has a High Ambition

South Korea's industry for advanced batteries—which are characterized by an extended life, faster recharging, and high-power levels—was $20 billion in 2021 (Yoon, 2023). South Korea's gross domestic product (GDP) was $1.8 trillion in 2021, which means that battery production accounted for slightly more than 1 percent of the ROK's GDP. Although battery manufacturing is a large and fast-growing industry in South Korea, it is not a dominant presence in the ROK economy yet in comparison with another key sector, semiconductors, which accounts for 8 percent of the country's GDP (O'Neill, 2023; Statistics Korea [e-나라지표], 2021).

However, South Korean companies are some of the world's largest vehicle battery manufacturers. LG Chem's LG Energy Solution is the world's second largest; Samsung SDI and SK On are also major players in the global

[11] In this context, the term *legacy* refers to logic semiconductors that belong to the 28-nm generation or older; a memory semiconductor with a half-pitch greater than 18 nm for DRAM or less than 128 layers for NAND that does not utilize emerging memory technologies (National Institute of Standards and Technology, U.S. Department of Commerce, 2023).

market. Table A.1 shows the market share for the largest EV battery makers. Together, these three firms hold 26 percent of the global market share. In non-Chinese markets, the share is more than double, at 54 percent (SNE Research, 2022; Zhang, 2022).

The ROK government has stated an ambitious goal to make the country into the number one EV battery producer in the world by 2030. For instance, South Korea announced plans to build an industrial network and coordinate financing of more than $1.3 billion to support battery development through tax incentives, R&D, and capital investments in the coming years (Blue House, 2021).

**TABLE A.1**

**Global Market Share of EV Battery Makers, by Gigawatt Hours of Capacity, in First Half of 2022**

| Company | Headquarters | Market Share |
|---|---|---|
| CATL | China | 34.8% |
| LG Energy Solution | South Korea | 14.4% |
| BYD | China | 11.8% |
| Panasonic | Japan | 9.6% |
| SK On | South Korea | 6.5% |
| Samsung SDI | South Korea | 4.8% |
| CALB | China | 4.1% |
| Guoxuan High-Tech | China | 2.9% |
| Sunwoda | Japan | 1.5% |
| SVOLT Energy | China | 1.3% |
| Others | N/A | 8.2% |

SOURCE: Features data from SNE Research, 2022.

NOTE: N/A = not applicable.

## South Korea Depends Heavily on China for Raw Materials, Processed Materials, and Battery Components

Critical raw materials for batteries—such as lithium, cobalt, graphite, and nickel—come from a variety of places around the world, including China, the Democratic Republic of the Congo, Chile, and Australia. Lithium, in particular, is needed to produce modern EV batteries. However, the vast majority of South Korean imports for three of the critical raw materials used in EV batteries are from China—88 percent of its imported lithium, 73 percent of its imported cobalt, and 93 percent of its imported graphite come from the PRC (Herh, 2023).

China also tends to dominate the processing of these minerals, which need significant refining to sufficiently reach the quality that is required for EV batteries. China controls about 58 percent of the market for refined lithium; four of the world's five largest lithium-processing companies are in China. The PRC also supplies 66 percent of the market for refined cobalt (Wu and Huy, 2022). One of the key refined minerals is lithium carbonate, which is produced from brines near lithium deposits. Traditionally, lithium carbonate has been one of the key refined minerals and is produced from brines near lithium deposits. Since the late 2010s and early 2020s, newer battery chemistries have been driving up demand for lithium hydroxide, which is usually produced from hard-rock mining (Barrera, 2019; Bisley International, 2021). Regarding South Korea's imports of lithium hydroxide, 84 percent came from China in the first seven months of 2022 (Chea, 2022). The ROK also imports about three-quarters of its cobalt oxides, another key mineral in battery production, from the PRC (Observatory of Economic Complexity, undated).

South Korea also depends heavily on China for the battery components, including anodes and cathodes, which are produced from advanced engineering of processed raw materials, such as lithium hydroxide. China produces 85 percent of the world's battery anodes and 70 percent of the world's cathodes. South Korea imports nearly 85 percent of the anodes and 73 percent of its cathodes from China (International Energy Agency, 2022).

Chinese companies even occupy commanding positions in both the cobalt and nickel supply chains outside China (Searcey, Forsythe, and

Lipton, 2021; Wu and Huy, 2022). The costs of these minerals and other precursors constitute the largest share of the final cost of an EV battery.

Therefore, practically speaking in terms of supply chain vulnerability, the South Korean industry needs raw materials, processed materials, and components from China to make high-capacity, high-endurance batteries for EV and other advanced technologies. Even if China could not completely cut off South Korean battery and EV manufacturers from these supplies, increasing their costs for South Korean companies could be detrimental to these companies' ability to compete in global markets. Demand for batteries and other environmentally friendly technologies continues to skyrocket, and competition over these minerals is likely to intensify (Searcey, Forsythe, and Lipton, 2021; Wu and Huy, 2022).

## South Korean Dependence on China Creates Both Opportunities and Risks for South Korean Manufacturers

Chinese companies manufacture approximately 70 percent of the world's advanced batteries and are expected to continue to dominate manufacturing through 2030 (Willuhn, 2020). Batteries are an essential part of any electrical generation and distribution system requiring storage, but major growth is expected in EV batteries. There are a variety of reasons why EVs might become attractive transportation alternatives, and the expected market growth is not simply a matter of the expected environmental benefits (Mordor Intelligence, undated).

LG Energy Solutions, Samsung SDI, and SK On are among the world's largest battery manufacturers and hope to cash in on an expected boom in demand from EVs, but the South Korean market is far too small to support major investment; these firms must operate and sell in other markets, such as China, the United States, and Europe, to maintain competitiveness. To meet the soaring demand in the global market, South Korean firms are working to secure additional supplies abroad. However, these efforts may take time, and manufacturers may struggle to meet the demand for battery materials, especially in the short term (Son, 2022). Not only are many raw materials produced in China, but Chinese firms control much of their production around the world, such as Congolese production of cobalt or Indonesian production of nickel. This control over production in the supply chain could lead to problems for South Korean firms if these overseas Chinese companies gave preferential treatment to non-ROK customers. Nonetheless, this potential threat can also be viewed as an opportunity. If South Korean manufacturers, in concert with their stakeholders in the public and private sectors in the ROK, can manage to strengthen their relationship with China, they may experience a bright and prosperous road ahead in the EV battery industry.

# Abbreviations

| | |
|---|---|
| AI | artificial intelligence |
| ATP | assembly, testing, and packaging |
| C2 | command and control |
| CHIPS | Creating Helpful Incentives to Produce Semiconductors |
| CPU | central processing unit |
| DF-41 | Dongfeng-41 |
| DRAM | dynamic random-access memory |
| EDA | electronic design automation |
| EV | electric vehicle |
| fab | fabrication facility |
| Gb | gigabit |
| GB | gigabyte |
| IP | intellectual property |
| IRA | Inflation Reduction Act |
| NAND | not and |
| nm | nanometer |
| PLA | People's Liberation Army |
| PLAN | People's Liberation Army Navy |
| PRC | People's Republic of China |
| RAM | random-access memory |
| R&D | research and development |
| ROK | Republic of Korea |
| SME | semiconductor manufacturing equipment |
| SMIC | Semiconductor Manufacturing International Corporation |
| TSMC | Taiwan Semiconductor Manufacturing Company |
| YMTC | Yangtze Memory Technologies Corporation |

# References

*Unless otherwise indicated, the authors of this report provided the translations of bibliographic details for the non-English sources included in this report. To support conventions for alphabetizing, sources in Korean are introduced with and organized according to their English translations. The original rendering in Korean appears in brackets after the English translation.*

Agreement Between the Republic of the Philippines and the Government of the United States on Enhanced Defense Cooperation, signed at Manila, Philippines, April 28, 2014.

Allen, Gregory C., *Choking Off China's Access to the Future of AI: New U.S. Export Controls on AI and Semiconductors Mark a Transformation of U.S. Technology Competition with China*, Center for Strategic and International Studies, October 11, 2022.

Alsop, Thomas, "Share of Integrated Circuit (IC), Integrated Device (IDM), and Fabless Company Sales in 2021, by HQ Location," webpage, Statista, January 26, 2023a. As of April 5, 2023:
https://www.statista.com/statistics/1052972/
ic-idm-and-fabless-sales-share-by-headquarter-location-of-company/

Alsop, Thomas, "DRAM Manufacturers Revenue Share Worldwide from 2011 to 2022, by Quarter," webpage, Statista, February 8, 2023b. As of March 28, 2023:
https://www.statista.com/statistics/271726/
global-market-share-held-by-dram-chip-vendors-since-2010/

Alsop, Thomas, "NAND Flash Manufacturers Revenue Share Worldwide from 2011 to 2022, by Quarter," webpage, Statista, February 9, 2023c. As of March 28, 2023:
https://www.statista.com/statistics/275886/
market-share-held-by-leading-nand-flash-memory-manufacturers-worldwide/

Alsop, Thomas, "Leading Semiconductor Foundries Revenue Share Worldwide from 2019 to 2022, by Quarter," webpage, Statista, March 14, 2023d. As of March 29, 2023:
https://www.statista.com/statistics/867223/
worldwide-semiconductor-foundries-by-market-share/

ASML, "A Backgrounder on Extreme Ultraviolet (EUV) Lithography," *Medium*, January 18, 2017.

Barbe, Andre, and Will Hunt, *Preserving the Chokepoints: Reducing the Risks of Offshoring Among U.S. Semiconductor Manufacturing Equipment Firms*, Center for Security and Emerging Technology, May 2022.

Barrera, Priscila, "Will Lithium Hydroxide Really Overtake Lithium Carbonate?" *Investing News*, June 27, 2019.

Bhattacharjee, Yudhijit, "The Daring Ruse That Exposed China's Campaign to Steal American Secrets," *New York Times*, March 7, 2023.

Bisley International, "What Is the Difference Between Lithium Carbonate and Lithium Hydroxide," webpage, February 14, 2021. As of December 2, 2022: https://bisleyinternational.com/what-is-the-difference-between-lithium-carbonate-lithium-hydroxide

Blue House, "President Moon: Goal to Be the 'No. 1 Battery Country in Name and Reality' by 2030" ["문 대통령 2030년까지 '명실상부한 배터리 1등 국가' 목표"], Republic of Korea Policy Briefing [대한민국 정책브리핑], July 9, 2021.

Brown, Peter, "TSMC Enters Volume Production with 3 nm Process Technology," *Electronics 360*, December 30, 2022.

Bureau of Industry and Security, U.S. Department of Commerce, "Addition of Huawei Non-U.S. Affiliates to the Entity List, the Removal of Temporary General License, and Amendments to General Prohibition Three (Foreign-Produced Direct Product Rule)," *Federal Register*, Vol. 85, No. 162, August 20, 2020.

Chea, Sarah, "After U.S. Law, Battery Makers Have Big Sourcing Problems," *Korea JoongAng Daily*, September 4, 2022.

Chiang, Sheila, "The World's Two Largest Memory Chipmakers Signal Demand Weakness May Have Bottomed Out," CNBC, November 6, 2023.

Choi, Jasmine, "NVIDIA Partners with SK Hynix; Set to Lead Next-Gen DRAM Market," *BusinessKorea*, June 15, 2023.

Cimpanu, Catalin, "FBI Is Investigating More Than 1,000 Cases of Chinese Theft of US Technology," *ZDNET*, February 8, 2020.

Clarke, Peter, "Report: Arm Proposes Change to IP Royalty Model," *eeNews Analog*, March 24, 2023.

Deutsch, Jillian, Eric Martin, Ian King, and Debby Wu, "US Wants Dutch Supplier to Stop Selling Chipmaking Gear to China," Bloomberg News, last updated July 6, 2022.

Flicker, Scott M., David S. Wang, Jia Yan, and David Cao, "Magnachip CFIUS Case Underscores Focus of U.S. Government on Semiconductor Supply Chain Security," Paul Hastings, September 17, 2021.

Gargeyas, Arjun, "The Chip 4 Alliance Might Work on Paper, but Problems Will Persist," *The Diplomat*, August 25, 2022.

Gross, Anna, "ASML Reveals Intellectual Property Theft by China Employee," *Financial Times*, February 15, 2023.

Henry, Ryan, Steven Berner, and David A. Shlapak, *Serious Analytical Gaming: The 360° Game for Multidimensional Analysis of Complex Problems*, RAND Corporation, RR-1764-OSD, 2017. As of March 18, 2024: https://www.rand.org/pubs/research_reports/RR1764.html

Herh, Michael, "Three Korean Battery Makers to Invest up to KRW20tn This Year," *BusinessKorea*, February 13, 2023.

Hoskins, Peter, "China Bans Major Chip Maker Micron from Key Infrastructure Projects," BBC, May 22, 2023.

Hruska, Joel, "How Are Process Nodes Defined?" *Extreme Tech*, February 12, 2021.

Hurst, Daniel, and Julian Borger, "Aukus: Nuclear Submarines Deal Will Cost Australia up to $368bn," *The Guardian*, March 13, 2023.

Hwang, Jeong-Soo, "Samsung's 3 nm Chip Production Yield Rises Sharply," *Korea Economic Daily*, January 9, 2023.

IC Insights, "Research Bulletin: Top Three Suppliers Held 94% of 2021 DRAM Marketshare," May 24, 2022.

Intel, "How Many Manufacturing Fabs Does Intel Have?" webpage, October 12, 2023. As of June 18, 2024: https://www.intel.com/content/www/us/en/support/articles/000089875/programs/intel-corporation.html

International Energy Agency, *Global Supply Chains of EV Batteries*, July 2022.

Jobst, Nina, "Value of Goods Exported from South Korea to China from 2000 to 2022," webpage, Statista, February 20, 2023. As of July 10, 2024: https://www.statista.com/statistics/657617/south-korea-exported-goods-value-to-china-since-free-trade-agreement/

Jobst, Nina, "Trade Balance of South Korea from 1991 to 2023," webpage, Statista, July 3, 2024. As of July 10, 2024: https://www.statista.com/statistics/934796/south-korea-trade-balance/

Johnstone, Christopher B., "Japan's Transformational National Security Strategy," Center for Strategic and International Studies, December 8, 2022.

Khan, Saif M., Dahlia Peterson, and Alexander Mann, *The Semiconductor Supply Chain: Assessing National Competitiveness*, Center for Security and Emerging Technology, January 2021.

Kim, Eun-jin, "Samsung's 2nd NAND Flash Plant in Xian Goes Live," *BusinessKorea*, April 1, 2022.

Kirton, David, and Max A. Cherney, "Huawei's New Chip Breakthrough Likely to Trigger Closer US Scrutiny, Analysts Say," Reuters, September 6, 2023.

Lee, Ji-yoon, "SK Chairman Says Corporate Survival Is Key amid U.S.-China Tensions," *Korea Herald*, September 22, 2022.

Lewis, James A., *Learning the Superior Techniques of the Barbarians: China's Pursuit of Semiconductor Independence*, Center for Strategic and International Studies, January 2019.

Mellor, Chris, "Samsung to Mass-Produce 1 Terabit Flash Chip," *Blocks & Files*, November 7, 2022.

Miller, Chris, *Chip War: The Fight for the World's Most Critical Technology*, Scribner, 2022.

Mordor Intelligence, *Electric Vehicles Market Size & Share Analysis—Growth Trends & Forecasts up to 2030*, undated.

Nam, Kyung-don, "Samsung, SK Hynix Lead Global NAND Flash Sales in Q2: Report," *Korea Herald*, September 26, 2022.

National Institute of Standards and Technology, U.S. Department of Commerce, "Preventing the Improper Use of CHIPS Act Funding," *Federal Register*, Vol. 88, No. 56, March 23, 2023.

Observatory of Economic Complexity, "Cobalt Oxides and Hydroxides," webpage, undated. As of December 3, 2022:
https://oec.world/en/profile/hs/cobalt-oxides-and-hydroxides

O'Neill, Aaron, "South Korea: Distribution of Gross Domestic Product (GDP) Across Economic Sectors from 2012 to 2022," webpage, Statista, October 19, 2023. As of December 30, 2023:
https://www.statista.com/statistics/375580/south-korea-gdp-distribution-across-economic-sectors/

Park, Kate, "Samsung Electronics Starts 3-Nanometer Chip Production Ahead of TSMC," *Tech Crunch+*, June 30, 2022.

Picarsic, Nathan, and Emily de La Bruyere, "American Semiconductor Leadership Suffers from Bad Defense, No Offense," *Newsweek*, September 28, 2022.

Platzer, Michaela D., Emily G. Blevins, and Karen M. Sutter, *Semiconductors: U.S. Industry, Global Competition, and Federal Policy*, Congressional Research Service, R46581, October 26, 2020.

Public Law 117-167, The CHIPS and Science Act, August 9, 2022.

Public Law 117-169, The Inflation Reduction Act, August 16, 2022.

"R&D Spending Dominated by U.S. Chip Manufacturers," *Dataweek*, August 31, 2022. As of April 30, 2023: https://www.dataweek.co.za/17236r

Saxena, Vishakha, "US 'to Extend' China Chip Curb Waivers for South Korea, Taiwan," *Asia Financial*, June 13, 2023.

Searcey, Dionne, Michael Forsythe, and Eric Lipton, "A Power Struggle over Cobalt Rattles the Clean Energy Revolution," *New York Times*, updated December 7, 2021.

Semiconductor Industry Association, *2022 Factbook*, May 2022.

"SK Hynix Says Has Developed Its Most Advanced 238-Layer Storage Chip," Reuters, August 2, 2022.

SNE Research, "2022 1H Global EV & Battery Performance Review," press release, July 18, 2022.

Snyder, Scott, and See-Won Byun, "S Korean Trade, Diplomacy Trending Away from China," *Asia Times*, March 28, 2024.

Son, Ji-hyoung, "Korean Battery Makers Set for Growth as Demand Outside China Soars," *Korea Herald*, September 20, 2022.

Song, Hyung-Suk, "SK Hynix Unveils Industry's Highest-Performing DRAM Chip," *Korea Economic Daily*, October 21, 2021.

"South Korea Exports Suffer Longest Losing Streak in Three Years, China Sales Tumble," Reuters, April 30, 2023.

Statistics Korea [e-나라지표], Semiconductors Industry Trend [반도체, 디스플레이산업 동향], 2021. As of December 12, 2022: https://www.index.go.kr/unity/potal/main/EachDtlPageDetail.do;jsessionid=t447-ZkzzndkiytvMoJEFDyCy2KusiPs-s4NJKg7.node11?idx_cd=A0002

Swanson, Ana, "Netherlands and Japan Said to Join U.S. in Curbing Chip Technology Sent to China," *New York Times*, last updated January 30, 2023.

Taiwan Semiconductor Manufacturing Corporation, "TSMC Fabs," webpage, undated. As of April 6, 2023: https://www.tsmc.com/english/aboutTSMC/TSMC_Fabs

Ting-Fang, Cheng, Lauly Li, and Yifan Yu, "Apple Freezes Plan to Use China's YMTC Chips amid Political Pressure," *Nikkei Asia*, October 17, 2022.

TSMC—*See* Taiwan Semiconductor Manufacturing Corporation.

U.S. Census Bureau, "Trade in Goods with China," webpage, undated. As of July 15, 2024: https://www.census.gov/foreign-trade/balance/c5700.html

"U.S. Chipmaker Magnachip, China's Wise Road End $1.4bln Merger Deal," Reuters, December 13, 2021.

U.S. Department of Commerce, "Department of Commerce Announces the Addition of Huawei Technologies Co. Ltd. to the Entity List," press release, May 15, 2019.

U.S. Department of Commerce, "Commerce Adds China's SMIC to the Entity List, Restricting Access to Key Enabling U.S. Technology," press release, December 18, 2020.

U.S. Department of Commerce, "Commerce Department Outlines Proposed National Security Guardrails for CHIPS for America Incentives Program," press release, March 21, 2023.

U.S. Department of the Treasury, "Remarks by Secretary of the Treasury Janet L. Yellen on the U.S.-China Economic Relationship at Johns Hopkins School of Advanced International Studies," press release, April 20, 2023.

"US Tech Curbs Could Halve Growth of China's Top Chipmaker," Bloomberg News, October 10, 2022.

Varas, Antonia, Raj Varadarajan, Jimmy Goodrich, and Falan Yinug, *Government Incentives and US Competitiveness in Semiconductor Manufacturing*, Boston Consulting Group and the Semiconductor Industry Association, September 2020.

Varas, Antonia, Raj Varadarajan, Jimmy Goodrich, and Falan Yinug, *Strengthening the Global Semiconductor Supply Chain in an Uncertain Era*, Boston Consulting Group and the Semiconductor Industry Association, April 2021.

Willuhn, Marian, "Battery Manufacturing Capacity Set to Increase Fourfold to 1.3 TWh by 2030," *PV Magazine*, August 12, 2020.

Wilson, Jason R., "Development of Intel's 1.8nm & 2nm Process Nodes Completed, Expected First Half of 2024," *Wccftech*, March 8, 2023.

Wu, Yu-Hsuan, and Phuong Quoc Huy, "Geopolitics of Critical Minerals," in *Critical Minerals: Global Supply Chains and Indo-Pacific Geopolitics*, National Bureau of Asian Research, Special Report No. 102, December 2022.

Yoon, John, "The Rechargeable Battery Industry in South Korea Starting to Charge Up," *Seoul Z*, August 7, 2023.

Zhang, Phate, "CATL Holds 34.8% of Global Power Battery Market Share in H1," *CnEVPost*, August 2, 2022.